P9-EDF-434

SAVE YOUR LIFE

The Executive's Guide to Personal Fire Safety

by Michael F. Laffey

With a special note on insurance
by Fred Mansbach, All Risk Associates, Ltd.

Executive Enterprises Publications Co., Inc.

ISBN 0-917386-66-3

Library of Congress Catalogue Card No. 81-66130

Copyright © 1981 by Executive Enterprises Publications Co., Inc.
33 West 60th Street
New York, New York 10023

CONTENTS

1 HOW TO SURVIVE A HOTEL FIRE   9

   HOTEL SITE PROFILE   11
     Prereservation Hotel Check   11
     At the Hotel   11
     On Your Floor   14
      Emergency Fire Exits   14
     Safety In Your Room   14
      Fire Alarm Systems   15
      Flashlight and Key   15
      Windows   16
      Bathroom Exhaust   17

   FIRE CONDITIONS   17
     A Fire in Your Room   18
     Fire Outside Your Room   20

   DO NOT PANIC   22

   EXECUTIVE PORTABLE SURVIVAL KIT   22

2 OFFICE BUILDING EMERGENCIES   25

HOW TO SURVIVE AN OFFICE BUILDING FIRE

   LIFE SUPPORT SYSTEMS   27
     Safety Checks   27
     Evacuation Procedures   28
     Smoke Detectors   30
     Fire Drills   30
     Elevators—Do Not Use!   32

   BOMB THREATS   33

   BUILDING POWER FAILURE   33

   SUGGESTED FIRE SAFETY PLAN FOR
   YOUR OFFICE BUILDING   35

   APPENDIX   36

3 CONFERENCE CENTERS   51

ATTENDING A SEMINAR OR BUSINESS MEETING   53
    Floor Diagram/Exit System Guide   53
    Emergency Evacuation Procedures   54

SELECTING A CONFERENCE CENTER
OFF-SITE SEMINAR OR MEETING   54
    Recommended Building Site Profile   54
    Automatic Sprinkler System   54
    Smoke Detectors   55
    Standpipe System   55
    Fire Communications System   55
    Emergency Lighting   56
    Portable Fire Extinguishers   56
    Fire Brigade   56
    Fire Evacuation Announcement   57

4 PLACES OF PUBLIC ASSEMBLY   61

RESTAURANTS   63
THEATRES AND CINEMAS   65
    Seating Arrangement and Stage Designs   66
    Fire Alarm System   66
    Exit Systems   66

5 HOME SAFETY   69

SURVIVAL TIPS   71

SMOKE DETECTORS   72

FIRST-AID FIRE EXTINGUISHING APPLIANCES   74
    Fire Extinguisher   74
    Garden Hose   75

6 INSURANCE   77
by Fred Mansbach

INTRODUCTION

Below are two examples, imagined and real, of fire emergency situations. As you read these descriptions think of how you would react if your life were to be threatened by a major fire.

HIGH-RISE HOTEL

You have decided to attend the annual conference of your professional organization in American City. The hotel has assigned you luxurious accommodations on the twenty-third floor. You are delighted with the magnificent view of the city's towers and spires. However, the first night at the hotel you are awakened from a sound sleep by the realization that your life is in danger. Sirens are screaming in the night; black smoke is creeping under your door, threatening to envelop the entire room; flames on your windowsill are blocking the panoramic view. A fire alarm has been ringing in the hall. You have awakened to the realization that a fire is raging out of control in your hotel.

If you panic and think there is nothing you can do, you will probably be a fire victim.

HIGH-RISE OFFICE BUILDING

At 10 a.m. company executives and their staffs at 919 Third Avenue in New York City, found their morning routine suddenly interrupted. In a matter of minutes, the fifty-story office building was engulfed by an out-of-control fire. The waves of black smoke cut off many fire exits and stairwells. The flames warned of a disaster that could take human lives by the hundreds. Panic-stricken occupants began breaking windows and running wildly in the corridors. Thousands of soot-covered employees groped their way down the fire stairs and stood dazed and coatless in the street watching the building burn under the cold, grey sky.

After the fire department extinguished the blaze, officials found three persons dead. Many others had suffered from smoke inhalation or sustained other injuries. The promptness and heroics of the New York Fire Department saved many lives.

Unfortunately, few people are prepared to respond calmly, efficiently, and knowledgeably should a fire endanger the building where they work, a hotel they visit on a business trip, or their homes.

Save Your Life is based on my twenty-five years of professional fire service. It is designed to inform you of the basic steps to be taken in a fire emergency. I have shared with you my knowledge of fire prevention and emergency action, hoping it can make the difference between saving and losing a precious life.

Michael F. Laffey

MICHAEL F. LAFFEY

Michael F. Laffey, safety engineer of Citibank's High-Rise Fire Management Program for the past five years, is dedicated to the improvement of life support systems in today's urban environment. His twenty years of professional service as captain in the New York City Fire Department demonstrates his concern with fire safety. An authority on fire control management in high-rise building structures, Mike Laffey has designed numerous fire prevention programs for New York City, for which he has received a special commendation from the fire commissioner of the City of New York. His academic credentials include a degree in fire technology from City College of New York and a B.S. degree from St. Francis College. He is currently studying for a master's degree in Occupational Safety and Health at New York University. Mike Laffey is a member of the Society of Fire Protection Engineers, American Society of Safety Engineers, and the National Fire Protection Association.

Chapter 1

How to Survive a Hotel Fire

HOTEL SITE PROFILE

Prereservation Hotel Check

At the Hotel

On Your Floor

  Emergency Fire Exits

Safety in Your Room

  Fire Alarm Systems

  Key and Flashlight

  Windows

  Bathroom Exhaust

FIRE CONDITIONS

A Fire in Your Room

Fire Outside Your Room

DO NOT PANIC

EXECUTIVE PORTABLE SURVIVAL KIT

At the Las Vegas MGM Grand Hotel, eighty-four lives were lost to fire. The fire at the Stouffer's Inn in Westchester claimed the lives of twenty-six, many of whom were executives attending business meetings. Six lives were lost to the inferno at the luxurious Inn on the Park in Toronto; eight to the fire at the Las Vegas Hilton. These tragic statistics point to a critical need for a fire protection program for executives whose work requires travel away from home and office.

HOTEL SITE PROFILE

Prereservation Hotel Check

The first step in your away-from-home fire protection plan is verification of fire safety at the hotel where you plan to stay. You or your secretary should call the hotel to secure information about its safety record. The fire/safety survival checklist is designed to assist in the selection of an adequately fire-protected hotel. The hotel fire/safety survival checklist will also make you aware of the factors involved in fire protection.

One of the checklist questions that is part of your pre-reservation information-gathering is: What is the height of the hotel building? The maximum rescue capability of aerial fire trucks is presently between eight and ten floors. Therefore, you should not accept a hotel room above the ninth floor.

Additional checklist questions are: Does the hotel have a fire safety plan? Is the hotel protected by a sprinkler or smoke detection system? If the hotel does not have a minimum life support system consisting of either sprinklers or smoke detectors, find accommodations elsewhere at a safer hotel. The rest of the checklist is to be filled out upon arrival at the hotel.

At the Hotel

Upon arrival at the hotel, begin your on-site fire protection inquiry. Look at the outside to determine whether the building has setbacks. Most high-rise buildings are large at the base and narrow on top. The setbacks, or indentations, offer refuge areas and escape possibilities during a fire. Also talk to the bellhop, who is a good source of information about the hotel's safety record. He may be useful in verifying the checklist information you secured by phone.

Find out at the front desk if a floor schematic evacuation plan is available. It is usually located behind the entrance door to your room.) Review the plan to locate fire exit stairs. (Following is a typical exit diagram.) Do not hesitate to secure the assistance of a hotel representative in filling out any part of the checklist.

Check fire exits on your floor.

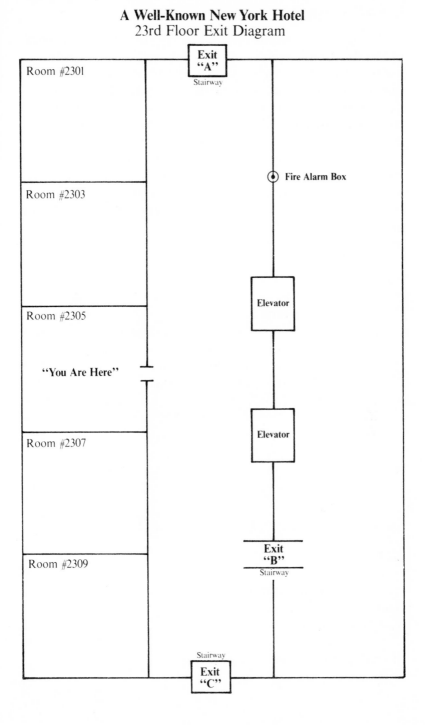

A Well-Known New York Hotel
23rd Floor Exit Diagram

On Your Floor

Emergency Fire Exits: It is imperative that you locate and inspect the fire exit stairs. Note the distance in feet you have to travel to all the available fire exit stairways, in all directions from your room, and enter this vital information on your fire/safety checklist. Count the number of doors you must pass before you enter the fire exit stairs. This safety step becomes very valuable when you are crawling down a smoky corridor. If an exit door is blocked or malfunctioning, the bellhop is obligated to report the situation immediately for repair. If you conduct your inspection alone, immediately report any defects to the hotel management. Check frequently to ascertain if repairs have been made.

Safety in Your Room

The first step to take when entering your room is to lay out your self-contained "survival kit." The kit, described in the last section of this chapter, contains a smoke detector, flashlight, and a hotel fire/safety checklist. Once the kit is set up, ascertain if the room contains a floor schematic evacuation plan indicating the fire exit stairs. If it does, review the plan with the bellhop. If no plan is present, request the assistance of your bellhop in filling out your personal hotel fire/safety survival checklist. It should contain the critical information listed here.

HOTEL FIRE/SAFETY SURVIVAL CHECKLIST

Name of Hotel _____

Address _____

Floor and Room Number _____

Approximate Height Aboveground _____

Telephone Numbers:

    Service Desk _____

    Fire Department _____

    Police Department _____

Fire Safety Plan _____

Number of exits available from your room to the
  fire stairs = Total _____

Exits in the *RIGHT* direction:
  Approximate number of feet _____
  Number of doors _____

Exits in the *LEFT* direction:
  Approximate number of feet: _____
  Number of doors _____

Does your room contain fire protection?
  Smoke Detector(s) _____
  Sprinkler System _____

Windows:
  Number in Room _____
  Sealed _____
  Sash _____

Bathroom Exhaust System _____

Is your "survival kit" set up? _____
  Smoke Detector _____
  Flashlight _____

Fire Alarm Systems: Note the location of the manual fire
alarm station on your floor. Review how it operates. If
instructions, which are usually printed on the cover of the
alarm box, are not available or you do not understand
them, ask for assistance.

Key and Flashlight: Develop the safe habit of placing
your key and flashlight in the same location every time
you stay in a hotel. It is recommended that you use a
nightstand for this purpose. The key and flashlight will be
close to your bed and will save you precious seconds look-
ing for them under emergency conditions. Practice grab-
bing for the key and light with your eyes closed. The point
being illustrated here is that if you are forced out of your
room under emergency conditions, you may have to return to it. For example, a heavy smoke or fire condition in

the corridor may cut off your escape to fire exit stairs, in which case you will need your flashlight at once.

Windows: Study the window of your room. Since most high-rise buildings have sealed windows, you will probably discover that your window cannot be opened. However, if you are staying in an older building or resort, your room may have a window that opens or a sliding glass door leading to a balcony. See if the balcony allows for access to another room or to a setback. This could be a lifesaver in an emergency situation. Look outside your window. Form a strong picture of the outside environment.

Place flashlight, door keys, copy of "Save Your Life" and smoke detector on night stand next to your bed.

Bathroom Exhaust: The exhaust system can aid in filtering the air in a room filling with smoke. If fire breaks out and smoke begins seeping into your room, turn on the bathroom exhaust. Keep it running for the duration of the emergency. It will alleviate the toxic effects of heavy smoke.

FIRE CONDITIONS

In an emergency situation, you may be subjected to two types of fire conditions: one inside your room, the other elsewhere in the hotel. These conditions require a different response.

Crawl under smoke cloud along hotel corridor.

A Fire in Your Room

Should you awake to fire or smoke in your room, take your key and flashlight from the nightstand, roll out of bed, and crawl to the door. Do not stand up! Heat and gas rise to the ceiling. By moving on your hands and knees you can better avoid inhaling the deadly gasses released during combustion. Carbon monoxide will render you unconscious in a matter of moments.

Before entering the corridor, touch the door and knob. If both are cool, slowly open the door and check the hallway for smoke. If the hall is clear, close the door behind you and activate the fire alarm on your floor. If there is a

Never use the elevator.

telephone in the hall, notify the front desk of the emergency condition.

On leaving a burning building, *never* use the elevator. Use the fire exit stairs. An elevator's call buttons, controls, and photoelectric light beams become unreliable when exposed to fire or smoke. Avoiding the elevator during a fire requires discipline, since leaving the building quickly is essential. But remember that elevator shafts, which extend the entire height of the building, serve as a flue for the fire's toxic smoke. After a killer fire, firefighters often find bodies in the elevator, when other inhabitants have escaped safely. The High-Rise Fire Safety Rules, Local Law #5 of New York City, were enacted as a result of

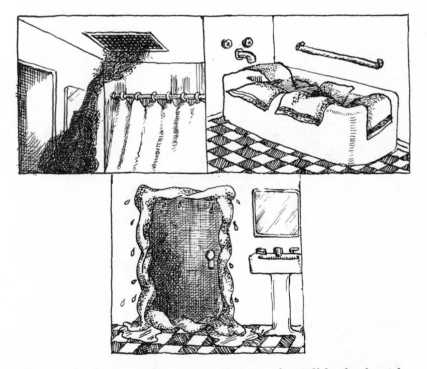

Turn on bathroom exhaust to remove smoke. Fill bathtub with cold water to soak towels. Stuff soaked towels under door to create seal.

devastating office building fires of the early 1970s, which resulted in the loss of five lives. All five victims of these fires were found in elevators.

Once you have reached the lobby via the fire exit stairs, notify the front desk. The management should be conducting a guest list verification.

Fire Outside Your Room

If a fire breaks out elsewhere in the hotel, the information you have recorded on your hotel fire/safety checklist will prove invaluable. If, on checking the knob and door for temperature, you find they are hot—DO NOT OPEN THE DOOR. Assume you are trapped in your room and take the following action:

Hang a sheet out of your window to signal firemen.

- Use the telephone numbers on your checklist to notify the front desk, fire department, and police department of your predicament.
- Turn on your bathroom exhaust to remove smoke.
- Fill your bathtub with cold water, soak towels, and stuff these under and around the door to your room to create a seal.
- If your windows can be opened, hang out a sheet or a bedspread as a distress signal to the fire department.
- If smoke begins seeping into your room, press a wet towel to your face and lay flat on the floor near an open window.

If the doorknob and door are cool, open the door a crack to see if the corridor is contaminated with smoke or fire.

Lay down near open window, and stuff soaked towels under door to create seal.

Be ready to close the door immediately. If the hall is only slightly filled with smoke, crawl on the floor (with flashlight and key), hugging the wall on the side where the fire exit is located. Smoke may obscure your vision. If you are on the wrong side of the hall, you could crawl past the exit. Count the number of doors to the exit. When you reach the exit, walk down the stairs holding firmly to the handrail.

DO NOT PANIC

If you hope to survive a fire, you must be committed to a safe evacuation plan. During a fire, the effects of shock and panic on human behavior are usually self-destructive. People often behave incorrectly during a fire because they have never been taught how to respond or they forget important instructions. When asked what they would do if their homes caught fire in the middle of the night, both children and adults too frequently answer they would hide under their beds or in closets. And this is exactly where firefighters find fire victims. Asked what they would do if trapped in a room filling with smoke, others say they would jump from windows. Instead, they should be sealing the crack under the door and yelling for help from an open window. In a fire, do not panic. Calmly and quickly put into effect the appropriate fire protection plan.

EXECUTIVE PORTABLE SURVIVAL KIT

Executive travelers should include in their luggage a self-contained portable survival kit. Included in the kit should be the following essential items:
- Ionization smoke detector
- Flashlight
- Hotel fire/safety survival checklist

The smoke detector serves as an early alert device. If it comes with a latch holding device, it should be placed on the top of a door. Otherwise place it on the tallest dresser. For maximum effectiveness, it should be at least six inches from the ceiling. If the ceiling is too low, place the detector on the dresser in your hotel room. Responding quickly to an emergency, the detector will sound an alarm if smoke enters the protected room. Before leaving on a trip, test the detector to make sure it is working. Include a spare battery in your kit.

As mentioned earlier, place the flashlight on your nightstand with your room key. If you need to evacuate your room, the light is essential to finding your way out of your room and into the corridor.

The executive who, on hearing the warning, "Fire," is able to carry out his or her own survival emergency plan is

Put smoke detector on room door.

very fortunate. By exhibiting clear thinking, becoming familiar with the hotel's layout and fire protection system, and traveling with the kit, you should be able to escape a fire danger zone.

FIRE DEATH RATES IN LARGE CITIES

Deaths per million persons, 1974-77 average. Excludes transportation fire deaths and deaths of nonresidents. *Source: National Center for Health Statistics.*

1	NEWARK	64.8	27	ROCHESTER	35.6
2	BIRMINGHAM, Ala.	59.8	28	WICHITA, Kan.	34.9
3	BOSTON	58.5	29	NEW ORLEANS	34.8
4	CLEVELAND	54.0	30	JACKSONVILLE, Fla.	33.4
5	LOUISVILLE, Ky.	53.6	31	KANSAS CITY, Mo.	32.2
6	BALTIMORE	53.1	32	AKRON, Ohio	31.8
7	ATLANTA	52.8		SEATTLE	31.8
8	PHILADELPHIA	49.9	33	NORFOLK, Va.	31.4
9	CHICAGO	49.5	34	CHARLOTTE, N.C.	30.2
10	BUFFALO	47.3		HOUSTON	30.2
11	WASHINGTON	46.3		OAKLAND, Calif.	30.2
12	JERSEY CITY	45.1	35	SACRAMENTO, Calif.	28.7
13	DETROIT	44.8	36	PHOENIX	28.2
14	PORTLAND, Ore.	44.1	37	MIAMI	26.7
15	SAN FRANCISCO	42.9	38	COLUMBUS, Ohio	26.1
	TULSA, Okla.	42.9	39	LOS ANGELES	25.9
16	MEMPHIS	42.7	40	LONG BEACH, Calif.	25.3
17	PITTSBURGH	42.5	41	TAMPA, Fla.	25.0
18	OKLAHOMA CITY	42.3	42	OMAHA	24.3
19	FORT WORTH	40.5	43	MILWAUKEE	23.6
20	INDIANAPOLIS	39.2	44	DENVER	22.7
21	CINCINNATI	38.7	45	ALBUQUERQUE, N.M.	22.4
	TOLEDO, Ohio	38.7	46	SAN ANTONIO	22.0
22	ST. PAUL	38.4		TUCSON, Ariz.	22.0
23	ST. LOUIS	36.7	47	SAN DIEGO	15.5
24	DALLAS	36.6	48	SAN JOSE, Calif.	14.8
25	MINNEAPOLIS	36.4	49	AUSTIN, Tex.	10.0
	NEW YORK CITY	36.4	50	EL PASO	9.1
26	NASHVILLE	35.8	51	HONOLULU	7.1

Chapter 2

Office Building Emergencies

How to Survive an Office Building Fire

LIFE SUPPORT SYSTEMS

Safety Checks

Evacuation Procedures

Smoke Detectors

Fire Drills

Elevators—Do Not Use

BOMB THREATS

BUILDING POWER FAILURE

Suggested Fire Safety Plan for Your Office Building

On June 23, 1980, at 7:39 p.m., an alarm signaled a fire threatening a modern forty-two-story office building at 299 Park Avenue in New York City. The fire, raging out of control on the twentieth floor, began spreading to the next floor. On the thirty-second floor of the building, lawyers were at work on the delicate legalities of the Chrysler Corporation's loan application to the federal government. Fortunately, the New York City Fire Department saved the lawyers' lives and the legal documents. Had this fire occurred three hours earlier during peak occupation of the building, the United States might have witnessed its first three-digit death toll in an office building fire.

Every executive who conducts business from a suite occupying the upper reaches of a high-rise building must recognize that the magnificent view of the urban landscape and maximum fire protection are often incompatible. The executive must ask a critical question: "How well am I protected in case of fire?" Aerial-ladder fire trucks can effect rescue operations only as high as the eighth or ninth floors of most buildings. The setback of a building can limit truck reach to even lower floors. Unless a high-rise is completely protected by a sprinkler system, the executive must recognize that his or her life is vulnerable to fire.

LIFE SUPPORT SYSTEMS

Safety Checks

Every executive concerned with fire safety should become familiar with his or her office building's fire exit system. You should be aware of all the fire exits on your floor. Make a simple diagram of halls and corridors from your office to the exit stairs. On the diagram note distances from your office to each exit. Scale your diagram to these distances. (Use the office exit diagram at the end of this chapter as a guide for your drawing.) Keep copies of the diagram with you at all times in your wallet or purse, briefcase, and desk drawer.

Periodically walk past each exit stairway. Check frequently that the entrances are passable and unlocked. Though security may be a problem in your building, fire exit stairs must always be accessible.

Evacuation Procedures

In the appendix you will find a guide to the organization of a fire safety brigade in your building. Once the alarm has been sounded, the best protection against life loss is an office staff that is prepared to meet the threat. Responsibilities must be assigned, teamwork encouraged, and discipline stressed as the key to successful fire evacuation.

An emergency evacuation order always comes as a surprise. Unlike lunch and coffee breaks, fires are insensitive to the important phone call, the crucial bid, and the last-minute deal. When you hear a fire alarm you should not need to be coaxed to leave your work. The alarm means your life is in danger.

Anticipating the fire conditons you may face will aid in deciding the correct evacuation procedure. The alarm itself will not inform you of the severity of the fire. Keep in mind that the smoke, not the flames, is the first hazard to life.

An alarm sounded for precautionary reasons will summon you to evacuate a hostile environment. Though your immediate area may not be threatened, you must maintain the discipline required to conduct a successful evacuation. At all times follow the direction of the floor's fire warden or the building management. You will probably be guided by those in charge to uncontaminated fire exit stairs. Remember to walk—and not run—down the stairs.

If you are evacuating an environment in which a hazy smoke has appeared at ceiling level, you must respond quickly. The smoke is an indication that the fire may rapidly intensify in your area. Do not hesitate. Walk briskly to and down the fire stairs.

The alarm may alert you to a heavy smoke condition threatening your immediate environment. Smoke often spreads very quickly in high-rise buildings. The burning of plastic (urethane foam) furniture yields a dense smoke. Additionally, the duct systems that provide welcome air conditioning during the summer and heat during the winter are a disservice in a fire. In the tragic Las Vegas MGM fire, the ducts and shafts pushed tremendous amounts of smoke to the upper floors. Over sixty of the total eighty-four deaths occurred between the nineteenth and twenty-third floors.

During an evacuation, your floor may undergo an instantaneous transition from a clear environment to a heavy smoke condition. You must not panic. The correct response is to crawl quickly to an uncontaminated stairway. Remember that smoke contains the toxic gas, carbon monoxide. Exposure to as little as a 1.3 percent content in the air causes death in a few minutes. Because it is lighter than air, carbon monoxide rises to the ceiling. By crawling, you will remain beneath the carbon monoxide smoke zone.

If feasible, bring the flashlight from your survival kit to an evacuation. Except when traveling, you should keep the kit in your office. Of course, if an alarm sounds when you are away from your desk, do not retrieve the flashlight.

An emergency evacuation requires discipline. Unfortunately, in our specialized modern age we are poorly trained to cope with fires. We have been taught to leave firefighting to the professionals; our natural instinct would be to run down the exit stairs in an emergency. To add to the problem, the exit stairs of many high-rise buildings have not been designed to handle evacuation of their total occupancy. The time needed to evacuate a high-rise is measured in hours, not minutes. Studies show that two hours and eleven minutes are required to empty a fifty-story building with one uncontaminated fire stairway and an occupancy of 240 persons per floor. In a forty-story building, under the same conditions, only twenty-six minutes are gained. For this reason the fire department often advises relocation to lower floors rather than building evacuation. Therefore, listen closely to all instructions from those in charge during an evacuation. Your area of refuge may be only three floors below your office.

Following are tips to ensure a successful evacuation of your floor and building should an emergency arise:

- Frequent fire drills should be put into practice to familiarize all personnel with the steps involved in orderly emergency exits.
- Exit directions and signs must be in place in corridors and doorways at all times.
- Your building must be protected with an alarm system to warn occupants and the fire department of a fire.
- Exits must be unobstructed, properly designed, and conveniently located to ensure that all high-rise occupants are accommodated in an evacuation.
- Evacuees must be aware of alternative exits, should the closest one be blocked by fire and smoke.
- Those in charge must control any panic that arises during an evacuation by giving clear and authoritative directions.

Smoke Detectors

Keep your "early alert" smoke detector from your survival kit on your desk or end table. Besides its obvious protection capability, the smoke detector will be a conversation piece—the best way of spreading the word to your staff and to visiting executives of the need for fire safety in office buildings.

Fire Drills

A formal definition of a fire drill is the method and practice of systematic, safe, and orderly evacuation of an area or building in the least possible time. Fire drills are absolutely essential to fire protection in high-rise structures. Practiced a minimum of four times a year, fire drills should involve all tenants in floor evacuation. Exits should be varied so that office personnel are familiar with the location of all exit stairs. Do not wait until an emergency situation to seek out alternatives to the exit closest to your office. Remember, the closest exit may be contaminated during an actual fire.

If your office enacts the guidelines in the appendix, drills will be conducted under the leadership of the appointed fire warden. His or her responsibilities include:
- Checking exits to make sure they are passable;
- Preventing panic and confusion during evacuation;
- Selecting evacuation routes;

Keep flashlight, smoke detector, exit diagram and copy of "Save Your Life" in office.

- Controlling traffic;
- Searching for stragglers;
- Counting occupants after an evacuation; and
- Organizing the return to the floor once it is safe.

The person in charge has the important responsibility of deciding when an evacuation should take place. When in doubt, always evacuate. Through fire drills, personnel will become familiar with the evacuation signal and the exit routes they should follow.

After a fire drill, the evacuation managers should meet to evaluate the success of the drill and to work out problems such as misunderstood directions or inefficient exit routes.

Stairways, Not Elevators

Do not use an elevator to escape a fire. Always evacuate via the fire exit stairs. Recent reports have indicated that elevators frequently travel to the fire floor, either by human or mechanical accident:

- The heat on the fire floor can deform and deteriorate the elevator call button in such a manner that a call is actually registered.
- The call buttons can be activated when exposed to smoke over a long period.
- A passenger, unaware of the fire, may press a call button for the fire floor.
- A passenger on the fire floor may press a down call button to escape the fire, but decide to use the nearest stairway.

The elevator, if it opens on the fire floor, can easily become a flue conducting the flames and smoke to other parts of the building. Another danger is that evacuees seeing an open elevator door may panic and forget that an elevator during a fire is an unreliable and frequently fatal means of escape. Evacuate a floor only by the fire stairs.

BOMB THREAT

The frustration of a bomb threat is that the building management never knows whether the warning is real or a prank. However, evacuation must be considered in any threat. The risk is too great to ignore the warning of a bomb or explosives hidden somewhere in the building. The building staff should never decide that a bomb threat is a false alarm since they would be legally liable if the threat turns out to be real.

When a building receives a bomb threat, the manager should immediately inform the police, fire department, and any other agency having jurisdiction over the emergency. By coordinating their efforts, these agencies will decide whether an evacuation is necessary and direct the activities of the building occupants. You, as a tenant, must follow their directives. You may be requested to leave the building or to relocate to another floor.

POWER FAILURE

Peak power demands, challenging the capacities of utility companies, frequently result in power failures. Some large office buildings are equipped with emergency generators that will keep the elevators running. In this case, evacuation can proceed quickly. Otherwise, personnel will have to leave the building via exit stairs. Keep your survival kit flashlight in repair and always have an extra supply of batteries. The flashlight will come in handy if you have to evacuate a building darkened by a power failure.

OTHER FIRE SAFETY IDEAS

• If your office is planning a move, make sure that the new building meets maximum fire safety codes.

Office Exit Diagram

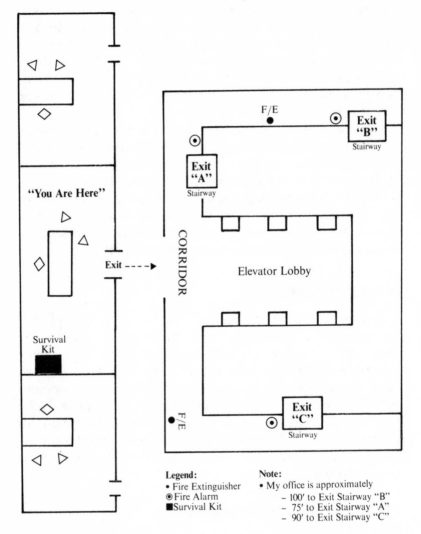

Legend:
- Fire Extinguisher
- ⊙ Fire Alarm
- ■ Survival Kit

Note:
- My office is approximately
  - 100' to Exit Stairway "B"
  - 75' to Exit Stairway "A"
  - 90' to Exit Stairway "C"

- Be vocal about the installation of a sprinkler system in your office building.
- Seek the advice of the fire department in the purchase of office supplies and furniture.
- Have an electrician check the wiring in your office. Alert your staff to the proper use of extension cords. Do not overload wires.

OFFICE BUILDING FIRE SAFETY PLAN

If the manager of your office building has not organized a fire safety plan, does not conduct fire drills, and has not selected a building emergency team or fire brigade, then it is up to you to alert him or her to the need for fire protection. Present to the manager the fire safety plan in this appendix. With modifications based on the needs of your building and the review of the local fire department, the plan can be instituted immediately. The fire safety plan will provide the fire protection required to maintain a safe and secure working environment.

Appendix

The following fire safety procedures, are intended for the occupants of high-rise buildings in the event of fire or an emergency. They offer an imaginative approach to fire protection which could be the difference between an orderly resolution of an emergency and a tragedy.

HIGH-RISE BUILDING SAFETY PROCEDURES

FIRE SAFETY PLAN

  Purpose and Objective

FIRE SAFETY DIRECTOR'S DUTIES

DEPUTY FIRE SAFETY DIRECTOR'S DUTIES

FIRE WARDENS AND DEPUTY FIRE WARDENS DUTIES

ORGANIZATION CHART FOR FIRE DRILL

BUILDING EVACUATION SUPERVISOR'S DUTIES

FIRE BRIGADE'S DUTIES

ORGANIZATION CHART FOR FIRE BRIGADE

OCCUPANTS' INSTRUCTIONS

FIRE PREVENTION AND FIRE PROTECTION PROGRAM

BUILDING INFORMATION FORM

FIRE COMMAND STATION DATA SHEET

OFFICE BUILDING FIRE PREVENTION EVALUATION PROGRAM

FIRE SAFETY PLAN

Building Address and Zip Code   Name of Building
Telephone Number

Purpose and Objective

Purpose

To establish a method of systematic, safe, and orderly evacuation of an area or building by its occupants in case of fire or other emergency, in the least possible time, to a safe area or by the nearest safe means of egress; also to establish the use of such available fire appliances (including sounding of alarms) as may have been provided to control or extinguish a fire and to safeguard human life.

Objective

To provide proper education through a written program for all occupants, to assure the prompt reporting of fire, the proper response to fire alarms, and the immediate initiation of fire safety procedures to safeguard life and contain fire until the arrival of the fire department.

A. Fire Safety Director's Duties

1. Be familiar with the written fire safety plan providing for fire drills and evacuation procedures.
2. Select qualified building service employees for a fire brigade and organize, train, and supervise such fire brigade.
3. Be responsible for the availability and state of readiness of the fire brigade.
4. Conduct fire and evacuation drills.
5. Be responsible for the designation and training of a fire warden for each floor and sufficient deputy fire wardens for each tenancy.
6. Be responsible for a daily check of the availability of the fire warden and deputy fire wardens and see

that up-to-date organization charts are posted.

7. Notify the owner or building manager when any individual is neglecting his or her responsibilities in the fire safety plan. The owner or building manager shall bring the matter to the attention of the firm employing the individual. If the firm fails to correct the condition, the fire department shall be notified by the owner or building manager.

8. In the event of a fire, report to the fire command station to supervise, provide for, and coordinate the following:
   (a) Notifying the fire department of the fire or fire alarm.
   (b) Operating of the fire command station.
   (c) Directing evacuation procedures as provided in the fire safety plan.
   (d) Reporting on conditions on fire floor to the fire department when it arrives.

9. Be responsible for the training and activities of the building evacuation supervisor.

B. Deputy Fire Safety Director's Duties

1. Follow the directives of the fire safety director.
2. Perform duties of fire safety director in his or her absence.

C. Fire Warden's and Deputy Fire Warden's Duties

1. The tenant or tenants of each floor shall, upon request of the owner or building manager, select responsible and dependable employees for designation by the fire safety director as fire wardens and deputy fire wardens.

2. Each floor of a building shall be under the direction of a fire warden for the evacuation of occupants in the event of fire. He or she shall be assisted by deputy fire wardens. A deputy fire warden shall be provided for each tenancy. When the floor area of a tenancy exceeds 7,500 square feet a

deputy warden shall be assigned for each 7,500 square feet or part thereof.

3. Each fire warden and deputy fire warden shall be familiar with the fire safety plan, the location of exits, and the location and operation of the fire alarm system.

4. In the event of fire or fire alarm, the fire warden shall ascertain location of the fire and direct the evacuation of the floor in accordance with instructions received and the following guidelines:

(a) The most critical areas for immediate evacuation are the fire floor and floors immediately above it. Evacuation from the other floors shall be instituted upon instruction from the fire command station or when conditions indicate such action. Evacuation should be via uncontaminated stairs. The fire warden shall avoid evacuation by stairs in use by the fire department. If this is not possible, he or she shall try to attract the attention of fire department personnel before opening the door to the fire floor.

(b) Evacuation to three or more levels below the fire floor is generally adequate. The fire warden shall keep the fire command station informed of his or her location.

(c) Fire wardens and their deputies shall see that all occupants are notified of the fire and that they immediately execute the fire safety plan.

(d) The fire warden on the fire floor shall, as soon as practicable, notify the fire command station of the particulars.

(e) Fire wardens on floors above the fire shall, after initiating the fire safety plan, notify the fire command station of the means being used for evacuation and any other particulars.

(f) The fire warden shall keep the fire command station informed of the means being employed for evacuation of occupants on his or her floor.

    (g) The fire warden shall determine that an alarm has been transmitted.

5. A chart designating employees and their assignments shall be prepared and conspicuously posted in each tenancy and on each floor of a tenancy that occupies more than one floor, and a copy shall be in the possession of the fire safety director. (See example of Organization Chart for Fire Drill and Evacuation Assignment on page 41.)

6. Fire wardens and building managers shall have available an updated list of all personnel with physical disabilities who cannot use stairs unaided. They shall make arrangements to have these occupants assisted in evacuation to three or more levels below fire floor and onto an uncontaminated floor.

7. Fire wardens shall be provided with identification during fire drills and fires, such as an armband.

8. Fire wardens are responsible for seeing that all persons on the floor are notified of a fire and are evacuated to safe areas. A search must be conducted in the lavatories to ensure that no one is in them. Personnel assigned as searchers can promptly and efficiently perform this duty.

9. Fire wardens shall check availability of applicable personnel on organization chart and provide for substitutes when positions on chart are not covered.

10. After evacuation, fire wardens shall perform a head count to ensure that all regular occupants known to have occupied the floor have been evacuated.

11. When alarm is received, the fire warden shall remain in the vicinity of the communication station on the floor to maintain communication with the fire command station and to receive and give instructions.

ORGANIZATION CHART FOR FIRE DRILL
AND EVACUATION ASSIGNMENT

BUILDING SAFETY DIRECTOR

_____

DEPUTY SAFETY DIRECTOR

_____

_____ FLOOR

FIRE WARDEN

_____

DEPUTY FIRE WARDENS

_____   _____

_____   _____

SEARCHERS

_____   _____

_____   _____

ALARM TRANSMISSION:

Any person discovering fire or smoke should without delay cause the transmission of a fire alarm by any of the following methods available:

1. Telephone;
2. Street alarm box;
3. Building fire alarms; if building fire alarm is not connected to central station, also notify fire department.

NOTE: Also notify fire and/or deputy fire wardens that alarm has been transmitted.

Date prepared _____   Date revised _____

D. Building Evacuation Supervisor's Duties

A building evacuation supervisor is required during nonworking hours when there are occupants in the building and when the fire safety director is off-duty. The building evacuation supervisor must:

1. Direct the evacuation of the occupants as provided by the fire safety plan.
2. During fire emergencies, operate the fire command station.

The building evacuation supervisor's training and related activities shall be under the direction of the fire safety director in accordance with these rules and the fire safety plan. Such activities shall be subject to fire department control.

E. Fire Brigade Duties

1. On receipt of a fire alarm, the fire brigade shall:

   (a) Report to the floor below the fire to assist in evacuation and provide information to the fire command station.

   (b) After evacuation of fire floor, endeavor to control spread of fire by closing doors.

   (c) Attempt to control the fire until arrival of the fire department, if the fire is small and conditions do not pose a personal threat.

   (d) Leave one member on the floor below the fire to direct the fire department to the fire location and to inform firefighters of conditions.

   (e) On arrival of the fire department, report to the fire command station for additional instructions.

   (f) Have a member designated as alarm box runner, who shall know the location of the nearest street fire alarm box and be instructed in its use. Such member shall immediately, upon receipt of information that there is a fire or evidence of fire,

ORGANIZATION CHART FOR FIRE BRIGADE

Names of Members
of Fire Brigade

Name and Title of
Person in Charge

_____   _____

_____   _____

_____

_____

_____

_____   Members assigned to assist
in evacuation

_____

_____

_____

_____   Members assigned to attempt to control small fires
_____   (minimum of two persons)

_____

_____   Alarm box runner for
transmitting alarms

_____   Back-up runner

_____   Member assigned to
communicate conditions to
fire command station

_____   Member assigned to floor
below fire to direct
fire department

Date prepared _____   Date revised _____

go to the street alarm box, transmit an alarm, and, upon the arrival of the fire department, direct it to the fire.

F. Occupants' Instructions

1. The appropriate parts of the approved fire safety plan shall be distributed to all tenants of the building by the building management when the fire safety plan has been approved by the fire commissioner.
2. The applicable parts of the approved fire safety plan shall then be distributed by the tenants to all their employees and by the building management to all their building employees.
3. All occupants of the building shall participate and cooperate in carrying out the provisions of the fire safety plan.

G. Fire Prevention and Fire Protection Program

1. A plan for periodic formal inspections of each floor area, including exit facilities, fire extinguishers, and housekeeping habits, shall be developed. A suggested plan is at the end of this appendix. The plan should include the following daily inspection checklist to be used by the fire safety director or other member of the fire control management team.

    (a) Check the availability of the fire warden and deputy fire warden.
    (b) Check each exit at the start of the day to determine that self-closing doors are in the closed position and not illegally locked in any manner.
    (c) See that there are no obstructions in corridors or aisle spaces.
    (d) See that necessary exit signs and lights, where required, are clearly visible and in good condition.

(e) Be sure that all personnel know the location and operation of fire extinguishers. The maintenance shall be supervised by the fire safety director.

(f) Check housekeeping (fire breeder). All areas shall avoid accumulation of combustible debris.

2. Provision shall be made for the monthly testing of communication and alarm systems.

H. Building Information Form

The fire safety plan shall include a building information form containing data as outlined in the following headings. Sufficient detail shall be entered after each of the headings to accurately portray the information.

Building Address_____   Zip Code _____

1. Owner or person in charge of building—name, address, and phone number.
2. Fire safety director and deputy fire safety director—name and phone number.
3. Certificate of occupancy. Location where posted, or duplicate attached.
4. Building height, area, class of construction.
5. Number, type, and location of fire stairs and/or fire towers.
6. Number, type, and location of horizontal exits or other areas of refuge.
7. Number, type, location, and operation of elevators and escalators.
8. Interior fire alarms, or alarms to central stations.
9. Communications systems, walkie-talkies, telephones, etc.
10. Standpipe system; size and location of risers, gravity or pressure tank, fire pump, location of siamese connections, name of employee with certificate number. Primary and secondary water supply, fire pump, and areas protected.

11. Sprinkler system; name of employee with certificate of fitness and certificate number. Primary and secondary water supply, fire pump, and areas protected.
12. Special extinguishing system, if any; components and operation.
13. Average number of persons normally employed in building—daytime and nighttime.
14. Number of handicapped people in building—location, daytime and nighttime.
15. Number of persons normally visiting the building—daytime and nighttime.
16. Service equipment such as:
    (a) Electric power; primary, auxiliary.
    (b) Lighting; normal, emergency, type and location.
    (c) Heating; type, fuel, location of heating unit.
    (d) Ventilation—with fixed windows, emergency means of exhaust of heat and smoke.
    (e) Air conditioning systems—brief description of system.
    (f) Refuse storage and disposal.
    (g) Firefighting equipment and appliances other than a standpipe and sprinkler systems.
    (h) Other pertinent building equipment.
17. Alterations and repair operations, if any, and the protective and preventive measures necessary to safeguard such operations with attention to torch operations.
18. Storage and use of flammable solids, liquids, and/or gasses.
19. Special occupancies in the building and the proper protection and maintenance thereof; places of public assembly, studios, and theatrical occupancies.

FIRE COMMAND STATION DATA SHEET

The following fire command station data sheet is to be kept at the fire command station and filled out by the fire safety director when a fire occurs in the building. This will ensure that all critical information is recorded for use by the arriving fire department.

1. What floor is alarm from? _____
2. Has alarm been transmitted to fire dept?
   Yes _____ No _____
3. Contact fire warden alarm floor. Ascertain conditions.
   _____ Minor fire
   (describe location, etc.) _____
   _____ More than minor fire
   (area, room, etc.) _____
   _____ Smoke condition—no fire apparent
   _____ Unknown
4. Is evacuation under way? _____
5. Any people reported trapped? _____
6. Contact fire warden, floor above alarm transmission. Ascertain conditions.
   _____ Fire
   (describe area, etc.) _____
   _____ Smoke only   Light _____ Heavy _____
   _____ No fire   _____ No smoke
7. Is evacuation under way? _____
8. Identification of elevators serving alarm floor
   Bank identification _____
   Elevator numbers _____
9. Sprinklers operating _____
   Number _____
   Location _____
10. Detectors activated _____   Number _____
11. Fans shut down? _____   Location _____
    Extent:
12. Number of occupants normally on floor at this time
    Alarm floor _____
    Floor above _____

OFFICE BUILDING FIRE PREVENTION EVALUATION PROGRAM

Daily Inspection Check-List

A. Check for the availability of the Fire Warden and Deputy Fire Warden

B. Check at the start of the day, each exit to determine that self-closing doors are in the closed position and not illegally locked in any manner.

C. No obstructions shall be permitted in corridors or aisle spaces.

D. Necessary exit signs and lights where required shall be lighted and in good condition.

E. Location and operation of fire extinguishers shall be known by all personnel. The maintenance shall be controlled by the Fire Safety Director.

F. Check housekeeping (fire breeder). All areas shall avoid accumulation of combustible debris.

Quarterly Inspection Checklist

A. Certificate of occupancy posted _____

B. Written record of the fire drills _____

C. Cellar or Subcellar:
   1. All areas clear of rubbish _____
   2. Boiler room _____

D. Housekeeping conditions (apply to entire building):
   1. All areas clear of rubbish _____
   2. "No Smoking" signs _____
   3. Aisle space, minimum 3 feet _____
   4. Correct number of extinguishers; proper inspection _____

E. Standpipe System:
   1. Siamese swivel serviceable _____
   2. Conditions at hose outlets satisfactory _____

F. Sprinkler System:
   1. City main shut off, labeled, and clear of obstruc-
      tions _____
   2. Control valves sealed open and labeled _____
   3. Access ladder provided for control valves _____
   4. Heads in place _____
   5. Clearance below sprinkler heads _____
   6. Written record of sprinkler tests _____

G. Roof Conditions:
   1. Water tanks in good condition _____
   2. Tank supports free of rust, no corrosion _____

H. Means of Egress:
   1. Hallways, stairways, passageways properly
      lighted _____
   2. Stairway clear of obstruction _____
   3. Steps, landings, handrails in good condition _____
   4. Exit signs _____
   5. Exit lights _____
   6. Doors self-closing and in good operating condi-
      tion _____

REMARKS:

Chapter 3

Conference Centers

ATTENDING A SEMINAR OR BUSINESS MEETING

Floor Diagram/Exit System Guide

Emergency Evacuation Procedures

SELECTING A CONFERENCE CENTER FOR AN OFF-SITE SEMINAR OR MEETING

Recommended Building Site Profile

Automatic Sprinkler System

Smoke Detectors

Standpipe System

Fire Communications System

Emergency Lighting

Portable Fire Extinguishers

Fire Brigade

Fire Department Protection

Fire Evacuation Announcement

The Stouffer's Inn fire in Westchester, New York, claimed the lives of twenty-six persons, half of whom were top executives attending a weekday meeting of a large Connecticut electronics firm. This tragedy has alerted meeting organizers to consider fire safety a primary criteria in the selection of a conference center.

Before attending a meeting or seminar, read the fire safety tips listed below. If you have the responsibility for arranging a meeting at a conference center, part of your duties should be to devise a maximum fire control management program for the attending executives. The checklist at the end of this chapter will be your guide in choosing a conference center that features the best fire protection available.

ATTENDING A CONFERENCE CENTER MEETING

Fire protection must be a priority consideration when attending a conference or seminar. Following is a guide to emergency preparation for the individual executive.

Floor Diagram—Exit System Guide

Arrive at the conference center at least forty-five minutes before the seminar is to begin. Walk through and study the center's exit system of corridors and stairs. Begin your tour with the conference room where the lecture will be held. Determine how many of the room's exits lead to a main corridor. From the corridor, walk to each exit that leads to the outside of the center. Draw a sketch of the conference room, corridors, and exits. This will be your "exit system guide," which you should have with you at all times at the center. At the end of this chapter you will find a sample exit system guide.

If you have difficulty understanding the layout of the building, seek assistance from the building management staff. If you find obstructions or locked exit doors, report these conditions immediately to the management.

Emergency Evacuation Procedures

At the first announcement of a fire emergency, you must react positively and quickly. Smoke and fire tend to spread rapidly in large meeting rooms. Do not pick up any books or clothing—you may need to keep your hands free to find your way to the exit. Move immediately to the exit, following the directional signs and your exit system guide to the appropriate fire exit door. If you enter a smoke contaminated area, crawl beneath the smoke zone, hugging the wall until you reach the exit and safe passage. Report to the fire command station where personnel will be conducting a verification list. Once directed to leave the building, you should not return under any circumstances.

SELECTING A CONFERENCE CENTER

As a meeting organizer, you have an important responsibility to your company and to those attending the meeting. Your first priority in selecting a conference center should be the safety of the building, not the beauty of its landscaping or interior decoration.

To assist the organizer, a checklist has been designed as a guide to selecting a conference center site with maximum fire protection for executives attending a meeting or seminar. (The checklist appears at the end of this chapter.) The required information can be obtained by telephone. However, you should visit the conference center and discuss the details of the checklist with the building's fire-control managers. The next section describes the minimum life support systems that should be present at the center.

Building Site Profile

Automatic Sprinkler System: A conference center that is not protected by an automatic sprinkler system is a maximum risk environment. The number of lives lost to fire reduced dramatically in buildings protected with sprinklers. In the past fifty years, the efficiency record of sprinkler systems is 96.2 percent in buildings threatened by fire. An automatic sprinkler system properly installed in the conference center you have selected is the best aid to life support for your group.

Smoke Detector Systems: Designed to detect the smoke that signals incipient combustion, these alarms, in tandem with a sprinkler system, are an excellent form of fire protection. While sprinklers fuse and activate at 165 degrees Fahrenheit, the smoke detector is designed to provide "early alert" protection. A warning from the smoke detector system can give the fire brigade time to put out the fire in its early stages with portable extinguishers before the overhead sprinkler system activates.

Standpipe System: Standpipe and hose systems allow for the manual application of water to interior fires. Though not a replacement for automatic fire extinguishers, standpipe systems should be present if automatic protection is not provided, especially in those areas of a building where hose lines will not reach from an outside fire hydrant. Standpipes are designed for fire department use to provide quick and convenient access to fire areas in large, low buildings or on the upper floors of high-rise buildings.

Fire Communications System: The conference center should be equipped with a fire command station located in the entrance floor lobby. The responding fire department will set up its command post here and work in liaison with the building's fire control management team to resolve the fire emergency. The ideal fire command station should be equipped with the following:

- Loudspeakers to broadcast on every floor, as well as in elevators and stair enclosures;

- Two-way voice communications with the fire warden stations on each floor, in the elevators, and the mechanical control center; and
- An alarm connected to the fire department.

Emergency Lighting: Find out if the conference center can arrange emergency lighting in the event of a power failure or an electrical breakdown during a fire. Evacuation from a burning building will be delayed if its occupants cannot find corridors and fire exits.

Portable Fire Extinguishers: Every conference center should be protected by portable fire extinguishers located throughout the building. Designed to put out small fires, extinguishers are a necessary and desirable complement to the fixed fire protection equipment. Extinguishers should be readily available throughout the center for use by those familiar with their operation.

Fire Brigade: The conference center should be protected by a fire brigade supervised by a fire safety director. Once a fire alarm has been transmitted, the fire brigade:
- Effects immediate evacuation of the fire floor by leading all occupants to safety;
- Controls the spread of fire and smoke by closing doors and isolating the condition; and
- Until the arrival of the fire department, attempts to control the fire while it is small and conditions do not pose a threat.

Fire Department Protection—Paid or Volunteer?
The meeting organizer should learn whether the local fire department is paid or volunteer. Eighty percent of United States fire departments are voluntary. The distinct advantages of a paid fire department are:
- Firefighters on continuous duty ensure a rapid response to a fire alarm;

- Fire prevention inspections of building sites are provided during business hours. The fire service can give valuable advice on improving fire protection and updating life support systems, and can check out the building exit systems.
- A familiarization program for all the firemen enables them to discuss the building fire safety plan with the fire control management team and the fire brigade.

Fire Evacuation Announcement

Once you have chosen a safe conference center, arrange for the building's fire safety director or other personnel to address your group on evacuation procedures and exit systems. This announcement should be made before the meeting begins. Many states have already incorporated into their fire code, evacuation announcements at public assemblies. A typical announcement might be:

> There are two exits from this meeting room. One exit is on your right and another is on your left. These exits lead into the main corridor, which will direct you to Exits A, B, and C. Please become familiar with these exits.

Include an exit system guide in the seminar packets or have a handout of the building's layout available before the meeting begins. In this way, the fire evacuation announcement will be reinforced with a visual aid.

You attend meetings and seminars to gain knowledge that will be profitable for you and your company. Applying the same principle to safety and fire protection, you must learn to recognize a fire-safe building.

If you are a meeting planner or seminar organizer, use the conference center selection guide checklist. If a seminar room is above the ninth floor and does not have a sprinkler system, smoke detectors, or a fire safety plan, disregard it as a conference center site. You will feel comfortable that you made a logical and intelligent decision.

Select a center that will offer maximum fire protection to you and your colleagues.

CONFERENCE CENTER SELECTION GUIDE CHECKLIST

Name of Conference Center _____

Location _____Phone _____

Building Height: _____

Floor of Conference Room: _____

Is the conference room within reach of the fire department's aerial apparatus? Yes _____ No _____

Construction:
  Fireproof material: _____
  Nonfireproof material: _____

Fire Control Management Program:

	Yes	No
Fire Safety Plan	_____	_____
Fire Safety Director	_____	_____
Fire Brigade	_____	_____

Fire Protection Systems:

	Yes	No
Sprinkler System	_____	_____
Smoke Detectors	_____	_____
Standpipe System	_____	_____
Fire Communications System	_____	_____
Emergency Power	_____	_____
Portable Fire Extinguisher	_____	_____
Fire Brigade	_____	_____
Fire Department Protection	_____	_____

Any recent fires? _____ _____

Conference Center Exit System Guide
(Fill Out Upon Arriving at Seminar or Meeting)

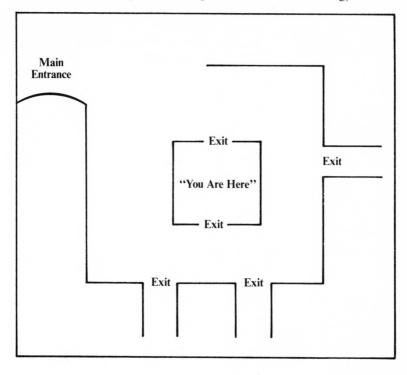

Chapter 4

Places of Public Assembly

RESTAURANTS

THEATRES AND CINEMAS

Seating Arrangement and Stage Designs

Fire Alarm System

Exit Systems

The tragic 1942 Coconut Grove nightclub fire in Boston claimed the lives of 492 people. More recently, in 1977, 165 revelers died at the Beverly Hills country club fire in Kentucky. Only after the death toll has been counted does national attention refocus on the importance of fire safety in places of assembly. Underlining the necessity of a safety code limiting occupancy in restaurants, bars, and clubs, the Coconut Grove fire illustrated what happens when too many people are gathered in a building constructed and decorated with highly flammable substances. The victims met their deaths in a fire that not only burned quickly but, in its consequences, was tragically severe. This chapter outlines life support tips for fire protection in places of public assembly.

RESTAURANTS AND NIGHTCLUBS

Before entering a restaurant or club, examine the building for fire exits. If possible, walk around the building. Inside, note all the available fire exits. Fire codes require restaurants and clubs to provide proper access to exits. Neither tables nor chairs are allowed to block exits. The most frequent offenders of safety codes are nightclubs that offer live entertainment on a full stage. Make sure that the exit path from your table is clear and unobstructed.

When seated at a restaurant, request a table that alllows a view of all available exits. Request a table away from the kitchen, as recent studies of restaurant fires indicate that 38 percent involved cooking equipment. Below is a typical floor diagram showing the preferred seating arrangement in a restaurant.

After you have placed your order, walk around the restaurant. Check the aisle space to the nearest exit. Make sure that the exit door works. Most restaurants are equipped with what is popularly known as "panic hardware." The releasing device on an exit door is designed to facilitate safe egress when a pressure of not more than fifteen pounds is applied to the door. Get acquainted with the exit door before you are forced to use it in an emergency. When you are comfortable with your seating arrangement, draw a simple sketch on your placemat of the layout of the room within the building and of the location of fire exits. This sketch will reinforce what you have learned during your walk through the restaurant.

The most spectacular of modern restaurants are those that revolve on top of high-rise buildings. Do not allow the

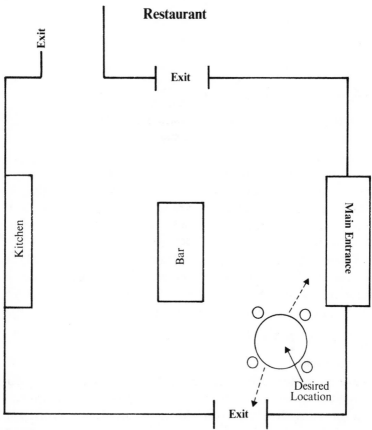

view from the top confuse your sense of the layout of the restaurant. The high-rise restaurant is usually required to have more exits than the hotel or office building below. Again, a simple sketch on your placemat is the best way for you to get your bearings.

In assessing whether a restaurant is safe, keep in mind that elevators, revolving doors, and fire escapes are not considered by the National Fire Protection Association Life Safety Code to be reliable fire exits. Revolving doors and fire escapes frequently have contributed to life loss in a fire. There have been instances where rusted fire escapes have collapsed with people on them during a fire department rescue. Also, persons have been fatally burned by fire breaking out of windows at a lower level. The sense of security they give, therefore, is a false one. As stated in previous chapters, do not use an elevator to escape a fire.

The most popular and elegant restaurants frequently are those which, night and day, offer their customers the ambience of a dimly lit room. However, semidarkness and unfamiliarity with the restaurant are factors that often lead to panic when an emergency situation occurs. If you are requested to evacuate a restaurant, leave immediately by the nearest uncontaminated fire exit. Do not attempt to claim your coat or other personal items. In the New York City Blue Angel nightclub fire in 1975, seven people who tried to retrieve their coats before fleeing the club, were overcome by smoke and died. Do not delay in evacuating a burning building.

THEATRES AND CINEMAS

In the early part of the century, 602 people died in the Chicago Iroquois Theatre fire. Subsequent fire codes and legislation protect the theatre patron from becoming a participant in a towering inferno scenario. However, when you go to the theatre, you should study the building's exit system.

Seating Arrangement and Stage Design: Theatres are designed to feature either continental, thrust, arena, or square seating. (See illustration at end of chapter.) The exit system varies with each design. If you arrive at the theatre twenty to thirty minutes before a performance, you will have time to walk through the theatre's exit system. Frequently, the playbill contains a schematic of the exit system, which you should study carefully.

Fire Alarms: Fire alarms in theatres are connected to the manager's office, dressing rooms, and auxiliary stage areas. The alarm is inaudible to the audience. Past incidents show that if a sudden shrill warning is sounded, crowds will spontaneously charge all the exits. To avoid panic in a fire emergency, the manager will calmly and deliberately direct the audience to evacuate the theatre. Fatalities will be avoided if the audience quickly and carefully follows these directives.

Exit Systems: Our five senses are integrated to inform us of our environment. When the environment poses aggravations, we are caused only temporary inconvenience. However, when our senses inform us that the environment is hostile, our involuntary reactions take over.

The planning of fire exit systems in theatres and cinemas takes into account these factors. People cannot be expected to behave logically in a stressful fire emergency. Because panic is contagious, the danger of irrational activity is much greater in a large crowd. Keep in mind that it is fear, rather than the actual fire, that causes panic. Fatal panics have occurred in safe buildings where people thought there was a fire. Conversely, when people have had confidence in a building and its exits, they have been able to execute an orderly evacuation, even though actual danger was present. When evacuating a theatre you must keep moving toward the designated safe exit. Any halt in the evacuation can touch off a panic response and exits will become blocked.

The concept of a self-supporting emergency evacuation program must begin with you. It is your only protection in places of public assembly where, in an unfamiliar environment, you may face potential danger with a large number of unknown people. Be prepared with a knowledge of the building and a willingness to follow the directions you are given.

Thrust Stage

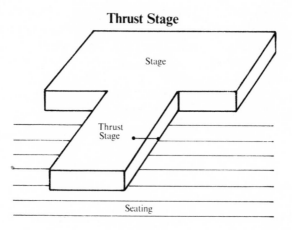

Square Stage

Stage

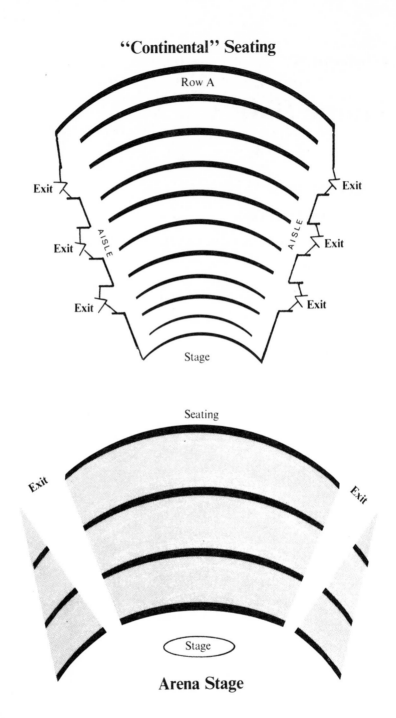

"Continental" Seating

Row A

Exit

Exit

AISLE

AISLE

Exit

Exit

Exit

Exit

Stage

Seating

Exit

Exit

Stage

Arena Stage

Chapter 5

Home Safety

SURVIVAL TIPS

SMOKE DETECTORS

FIRST-AID FIRE EXTINGUISHING APPLIANCES

Fire Extinguisher

Garden Hose

As a fire-conscious executive away on business, you may worry about how your family would cope should a fire start at home. To know that your family members and home are protected in a source of comfort. This can happen only if you institute a family emergency evacuation plan and install smoke detectors. You must provide your home are protected is a source of comfort. This can haping, with an "early alert" smoke detection system and an escape plan.

The installation of smoke detectors lulls many families into a false sense of security. Home protection is not complete unless the family has planned and practiced a thorough evacuation strategy. The smoke detector alarm must summon your family to automatic execution of an escape plan. Frequent drills will ensure that family members will respond quickly and with common sense to a fire. Otherwise your family may become part of the fire problem. If your family panics or becomes disoriented, the fire department's commitment to extinguishing the fire and saving lives becomes that much harder.

Every family member should be involved in planning fire evacuation drills. Establishing primary and secondary escape routes should involve family discussion of a properly executed evacuation program. Fire drills should be conducted at least four times a year. If family members change bedrooms or your home undergoes alterations or new construction, determine with your family if a new evacuation plan should be developed.

SURVIVAL TIPS

Each family member should be aware of the following survival tips:
- Bedroom doors should remain closed during sleeping hours. Closed doors afford protection from heat and smoke and allow extra escape time.

71

- Family members should become familiar with the sound of the smoke detector's alarm. They should know the alarm means immediate evacuation of the house.
- No one should waste time getting dressed or gathering valuables. Jewelry, money, favorite mementos, and even those pets must be left behind that don't evacuate immediately with the family.
- Touch the door before leaving a room. If the knob or the door is hot, do not open it. If feasible, use the window as an alternative exit. Family members who are on the first and second floors should be encouraged to drop from the window. Use a chair or other heavy object to break the window if it is jammed. The window frame should be cleared of jagged glass, and blankets placed over the sill to avoid dangerous cuts.
- Since smoke, heat, and fire rise to the ceiling, crawling is the safest mode of travel through a smoky area.
- Once out of the house, family members should meet at a prearranged rendezvous for a count.
- Call the fire department once outside the house. Do not call from your home.
- Remember that a seemingly small fire, confined to one part of a building, can suddenly spread because of flashover. Flashover is caused by thermal radiation feedback from the ceiling and upper walls, which have been heated by the fire. The radiation feedback gradually heats the contents of the fire area. When all the combustibles in the space are heated to their ignition temperatures, simultaneous ignition occurs.
- Once outside the house, stay out. Do not return. Many people lose their lives returning to a burning house to rescue a family pet or remove valuables.

SMOKE DETECTORS

The number of smoke detectors required for a home is based on the size and space configurations of the dwelling.

For instance, an apartment may require only one detector whereas a single-family house should be protected by one detector per landing. The following tips serve as a guide in the purchase and installation of your home smoke detector:

- Verify that the smoke detector has warranted approval from a reliable testing laboratory such as Factory Mutual or Underwriters Laboratory.
- Follow the manufacturer's recommendations and instructions for installing a smoke detector. It is a simple procedure.
- Place smoke detectors near sleeping areas, preferably in hallways or areas adjacent to bedrooms.
- Use the detector to protect your escape routes. Bed-

Plan an evacuation route and go to prearranged location.

rooms are usually located away from convenient exits. Place your detector in areas through which your family must pass during escape. The alarm will prevent your family from being trapped by dense smoke or flames blocking impassable escape routes.

- Place smoke detectors in the center of the ceiling of the designated room or corridor. This is the preferred location. If the center of the ceiling is obstructed, mount the detector at least six inches from the wall. Detectors that are mounted on the wall should be six to twelve inches from the ceiling.
- Place smoke detectors at the highest point on a sloped ceiling.
- Test the smoke detector before completing installation. Have all family members go to their bedrooms, close their doors, and listen for the alarm. Each person must be able to hear the alarm.
- Consult your local fire department if you have any questions about the installation procedures. Call if you need any assistance.

FIRST-AID FIRE EXTINGUISHING APPLIANCES

Fire Extinguisher

Most fires start small and can be easily extinguished if the proper type and appropriate amount of extinguishing agent is readily available and properly applied. Various classes and sizes of portable fire extinguishers are suitable for home use. Extinguishers are rated for use against either Class A fires, those involving ordinary combustibles such as paper or wood; Class B fires, involving flammable or combustible liquids; or Class C fires, involving electrical equipment. Extinguishers for all three classifications are available in various pound weights.

If you decide to maintain only one extinguisher, the multipurpose (A:B:C) dry chemical extinguisher is the recommended appliance. The manufacturer's label designates the purpose of the extinguisher and gives instructions for its use. The kitchen should have its own extinguisher and an additional one should be stored in the basement. A demonstration of how to operate the fire extinguisher should be part of family fire-safety planning. The extinguisher need not be discharged during the practice session. A dry run will suffice. Competent service companies are equipped to recharge extinguishers after they are used or if an inspection indicates such a need.

Garden Hose

A simple garden hose can also aid in fighting fires in the home. The hose should extend from a water hook-up in the basement, kitchen, or bathroom to the farthest potential fire area. Garden hose protection is particularly important for those living in secluded areas far from a local fire department.

Imagine arriving home from a business trip to the news that your family survived a perilous fire with no casualties. They owed their safety not to good luck, but to the installation of smoke detectors and the execution of a carefully worked-out evacuation plan. Statistics compiled by the National Fire Protection Association indicate that of the 12,000 deaths by fire recorded annually in the United States, two-thirds occur in resident dwellings. When you are away from home on business, you will want to know that your family is guarded by a fire protection program.

Chapter 6

Insurance

by Fred Mansbach

One of the most important—and sometimes overlooked—considerations for the frequent traveler is protection against personal injury and property loss. The need for protection in the form of insurance becomes particularly acute in the event of a fire at your hotel, for example, when you won't have time to think about collecting your belongings before evacuating your room.

Careful analysis and planning, then, can be your best defense against loss or damage. Just as a student traveler always checks and reconfirms travel arrangements (i.e. flight tickets, hotel reservations), you, as an executive, should give thought to the protection needed while away from home. This can easily be done even before you leave the house simply by considering the following:

- While traveling, am I fully and properly protected from injury from any cause?
- Is the property I'm taking with me fully protected from loss?

Under normal conditions, an individual traveler, whether self-employed or on a corporate trip, will be insured for accident, sickness, or death under the same program that covers him or her every other day of the year. Insurance companies do offer special "trip insurance" through brokers and agents and, for the last-minute shopper, they sell insurance at most airports. However, if you frequently find yourself buying these short-term travel policies, your basic insurance program needs reviewing.

Protection against injury due to accident or illness is customarily provided by one's employer and such protection should extend while you are traveling on company business. Loss of earnings as a result of injury or sickness is also a matter of company policy and, if necessary, can be supplemented by purchasing your own insurance to cover such contingencies. Insurance provided by your employer must be carefully read and understood and, with the aid of your insurance broker or agent, augmented to provide the fullest protection needed.

While your employer may provide you with protection while traveling, it is wrong to assume that such protections extends to persons traveling with you or to the personal property you may be carrying. The greatest risk by far to your personal property is loss by fire or theft. Remember, personal property encompasses everything from your shoes to the cash in your pocket, and you are subject to loss at any time, day or night, asleep or awake. Careful thought must be given to the risks and the methods to protect against those risks.

A serious hotel fire exposes you to a prime risk. Experience has shown that there is virtually no time to salvage one's personal belongings once an alarm has been sounded. The first thought, and rightfully so, is the saving of lives. But assume for the moment that you are involved in a hotel fire. You are aroused from a sound sleep and quickly and safely make your way to the street. Your property, at this point, can become the victim of the fire, attendant smoke damage, water damage, or even theft. Don't be misled into believing that the hotel is responsible for your loss. In some cases, as a gesture of goodwill and to forestall serious litigation, hotels will assume some or all of the property loss resulting from a fire. But they are by no means obliged to do so. Very often it requires a court of law to determine the hotel's responsibility—or lack of it.

Clearly, the most practical method to avoid the necessity of seeking legal recourse is to provide yourself with the proper insurance coverage. Your negotiations are then with your own broker or agent, who, in turn, would handle any claims against the hotel. In most states, protection for off-premises or away-from-home loss can be provided by an addition to your homeowners' or tenants' policy. For the frequent traveler, the need for extended protection is critical and the cost negligible. Pay particular attention to the limitations on such extensions with regard to jewelry, furs, cash, etc. These limitations can be modified to fit your needs. Here again, the services of a qualified broker or agent are a must.

Should you find yourself in a serious hotel fire involving personal injury and property loss beyond the scope of insurance coverage, it may become necessary to consult an attorney to determine what, if any, action can be taken. In the long run, however, proper insurance coverage is your best source of protection against financial loss.

Bibliography

Bryan, Dr. John L., AUTOMATIC SPRINKLER AND STANDPIPE SYSTEMS, National Fire Protection Association.

FIRE PROTECTION HANDBOOK, 14th Edition, National Fire Protection Association.

LIFE SAFETY CODE, National Fire Protection Association.

Phillips, Dr. Ann, "The Physiological and Psychological Effects of Fires in High Rise Buildings," Factory Mutual Record, May-June, 1973.

HOTEL FIRE/SAFETY SURVIVAL CHECKLIST

Name of Hotel _____

Address _____

Floor and Room Number _____

Approximate Height Aboveground _____

Telephone Numbers:
    Service Desk _____
    Fire Department _____
    Police Department _____

Fire Safety Plan _____

Number of exits available from your room to the
    fire stairs = Total _____

Exits in the *RIGHT* direction:
    Approximate number of feet _____
    Number of doors _____

Exits in the *LEFT* direction:
    Approximate number of feet: _____
    Number of doors _____

Does your room contain fire protection?
    Smoke Detector(s) _____
    Sprinkler System _____

Windows:
    Number in Room _____
    Sealed _____
    Sash _____

Bathroom Exhaust System _____

Is your "survival kit" set up? _____
    Smoke Detector _____
    Flashlight _____

HOTEL FIRE/SAFETY SURVIVAL CHECKLIST

Name of Hotel _____

Address _____

Floor and Room Number _____

Approximate Height Aboveground _____

Telephone Numbers:

    Service Desk _____

    Fire Department _____

    Police Department _____

Fire Safety Plan _____

Number of exits available from your room to the
    fire stairs = Total _____

Exits in the *RIGHT* direction:

    Approximate number of feet _____

    Number of doors _____

Exits in the *LEFT* direction:

    Approximate number of feet: _____

    Number of doors _____

Does your room contain fire protection?

    Smoke Detector(s) _____

    Sprinkler System _____

Windows:

    Number in Room _____

    Sealed _____

    Sash _____

Bathroom Exhaust System _____

Is your "survival kit" set up? _____

    Smoke Detector _____

    Flashlight _____

HOTEL FIRE/SAFETY SURVIVAL CHECKLIST

Name of Hotel _____

Address _____

Floor and Room Number _____

Approximate Height Aboveground _____

Telephone Numbers:
    Service Desk _____
    Fire Department _____
    Police Department _____

Fire Safety Plan _____

Number of exits available from your room to the
    fire stairs = Total _____

Exits in the *RIGHT* direction:
    Approximate number of feet _____
    Number of doors _____

Exits in the *LEFT* direction:
    Approximate number of feet: _____
    Number of doors _____

Does your room contain fire protection?
    Smoke Detector(s) _____
    Sprinkler System _____

Windows:
    Number in Room _____
    Sealed _____
    Sash _____

Bathroom Exhaust System _____

Is your "survival kit" set up? _____
    Smoke Detector _____
    Flashlight _____

HOTEL FIRE/SAFETY SURVIVAL CHECKLIST

Name of Hotel _____

Address _____

Floor and Room Number _____

Approximate Height Aboveground _____

Telephone Numbers:
    Service Desk _____
    Fire Department _____
    Police Department _____

Fire Safety Plan _____

Number of exits available from your room to the
    fire stairs = Total _____

Exits in the *RIGHT* direction:
    Approximate number of feet _____
    Number of doors _____

Exits in the *LEFT* direction:
    Approximate number of feet: _____
    Number of doors _____

Does your room contain fire protection?
    Smoke Detector(s) _____
    Sprinkler System _____

Windows:
    Number in Room _____
    Sealed _____
    Sash _____

Bathroom Exhaust System _____

Is your "survival kit" set up? _____
    Smoke Detector _____
    Flashlight _____

HOTEL FIRE/SAFETY SURVIVAL CHECKLIST

Name of Hotel _____

Address _____

Floor and Room Number _____

Approximate Height Aboveground _____

Telephone Numbers:

    Service Desk _____

    Fire Department _____

    Police Department _____

Fire Safety Plan _____

Number of exits available from your room to the
    fire stairs = Total _____

Exits in the *RIGHT* direction:

    Approximate number of feet _____

    Number of doors _____

Exits in the *LEFT* direction:

    Approximate number of feet: _____

    Number of doors _____

Does your room contain fire protection?

    Smoke Detector(s) _____

    Sprinkler System _____

Windows:

    Number in Room _____

    Sealed _____

    Sash _____

Bathroom Exhaust System _____

Is your "survival kit" set up? _____

    Smoke Detector _____

    Flashlight _____

HOTEL FIRE/SAFETY SURVIVAL CHECKLIST

Name of Hotel _____

Address _____

Floor and Room Number _____

Approximate Height Aboveground _____

Telephone Numbers:

    Service Desk _____

    Fire Department _____

    Police Department _____

Fire Safety Plan _____

Number of exits available from your room to the
    fire stairs = Total _____

Exits in the *RIGHT* direction:

    Approximate number of feet _____

    Number of doors _____

Exits in the *LEFT* direction:

    Approximate number of feet: _____

    Number of doors _____

Does your room contain fire protection?

    Smoke Detector(s) _____

    Sprinkler System _____

Windows:

    Number in Room _____

    Sealed _____

    Sash _____

Bathroom Exhaust System _____

Is your "survival kit" set up? _____

    Smoke Detector _____

    Flashlight _____

AN IMPORTANT MESSAGE TO OUR READERS

Now that you've had the opportunity to read about the many practical life-saving tips and procedures found in this book, consider sharing this valuable information with your friends, associates and family. There is no better way to show your concern for their safety than by giving them a copy of SAVE YOUR LIFE. If you have responsibility for a business or a large office staff, you may wish to order in larger quantities.

To receive additional information on prices and attractive discounts on quantity purchases, please send us the form (or information) found below. Thank you.

Please send to:

EXECUTIVE ENTERPRISES PUBLICATIONS CO.
Department: Fire Safety Books
33 West 60th Street
New York, N.Y. 10023

I am interested in:

☐ making SAVE YOUR LIFE available to my organization and staff.
Please send me your discount schedule for _____ quantity units
of SAVE YOUR LIFE. I understand there is no obligation.

☐ the latest catalog of books and related business seminars
from Executive Enterprises, Inc.

Name _____

Title _____

Firm/Organization _____

Street _____

City _____ State _____ Zip _____

Telephone (_____) _____. If you wish custom
imprinting information on your bulk purchase of SAVE YOUR LIFE.